Activity

Experiment: Build a Mini Solar Oven

Materials: A small cardboard box, aluminum foil, plastic wrap, black construction paper, and tape.

Instructions:

- Line the inside of the box with black paper (it absorbs heat).

- Cover the flap and inside of the box with aluminum foil to reflect sunlight.

- Place a snack inside (like a marshmallow) and cover the box with plastic wrap to trap the heat.

- Put the box in direct sunlight and watch as the sun's heat warms up your mini oven!

Learning Moment: This simple oven uses sunlight to create heat, just like solar panels use sunlight to make electricity

Wind Energy: Power from the Air

Wind is powerful! we see trees sway, kites fly, and waves crash. But did you know we can also use the wind to create electricity? Wind turbines are special machines that capture the wind's power.

When the wind blows, it spins the blades of the turbine. These spinning blades turn a part inside called a generator, which creates electricity. The faster the wind blows, the more electricity we can make!

What is Renewable Energy?

Non-Renewable Energy

Solar Energy: Power from the Sun

Wind Energy: Power from the Air

Hydropower: Energy from Water

Biomass: Energy from Nature's Leftovers

Geothermal Energy: Heat from the Earth

Tidal Energy: Power from the Ocean's Waves

Wave Energy: Power from Ocean Waves

What is Renewable Energy?

Renewable Energy is energy that comes from natural resources that never run out! This means we can use them over and over again without depleting them. Some of the most common sources of renewable energy include:

Sunlight: Solar energy uses sunlight to create electricity or heat.

Wind: Wind energy captures the power of the wind to turn turbines and generate electricity.

Water: Hydropower uses the movement of water, such as rivers or tides, to produce energy.

Heat from the Earth: Geothermal energy comes from the heat stored inside our planet.

Plants and Waste: Biomass energy uses organic materials like plants, wood, and even some types of waste to produce fuel.

Why is Renewable Energy Important?

Renewable energy is a cleaner way to power our lives. Unlike fossil fuels (like coal, oil, and natural gas) that can run out and produce pollution, renewable energy sources are eco-friendly and can help reduce harmful effects on the planet. As long as the sun shines, the wind blows, and rivers flow, we can keep using renewable energy!

Fun Fact:

Did you know that just one hour of sunlight could power the entire Earth for a whole year if we could capture it all? That's the amazing potential of renewable energy!

1. Solar Energy 🌞
2. Wind Energy
3. Hydropower
4. Biomass
5. Geothermal Energy
6. Tidal Energy
7. Wave Energy

Solar Energy: Power from the Sun

The sun shines brightly every day, giving us light and heat. But did you know we can also use the sun's energy to create electricity? Solar panels can capture sunlight and turn it into power we can use to light up our homes, run appliances, and even charge our gadgets!

When sunlight hits a solar panel, tiny pieces inside the panel (called solar cells) absorb the light. These cells then turn the sunlight into electricity we can use to power things in our homes and cities.

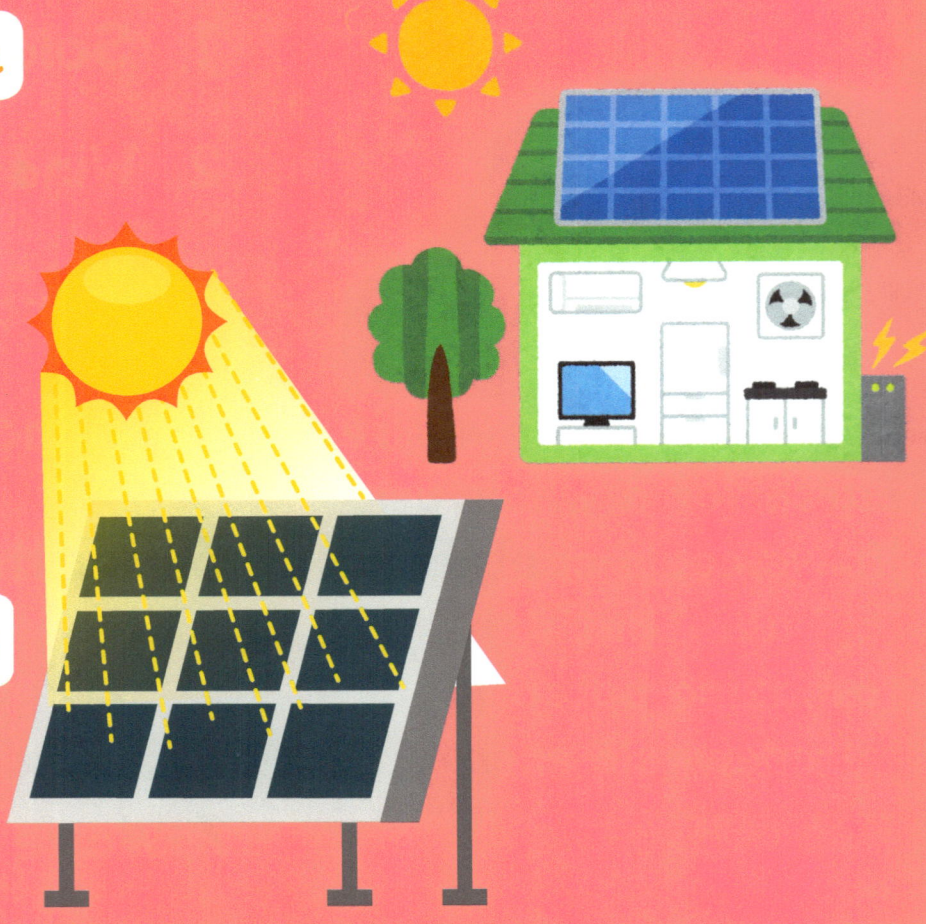

Activity

Make a Paper Pinwheel

Materials: Colored paper, a push pin, a pencil with an eraser, and scissors.

Instructions:

1. Cut a square piece of paper and fold it diagonally to make an X.

2. Cut along the lines partway, then fold the corners toward the center.

3. Secure with a push pin in the center, attaching it to the pencil eraser.

4. Hold the pencil and blow on the pinwheel to see it spin, just like a mini wind turbine!

Learning Moment: spinning pinwheel is similar to how a wind turbine works by catching the wind.

Hydropower: Energy from Water

Water is incredibly powerful! When it flows down rivers or falls from high places, it has a lot of energy. Hydropower uses this energy to make electricity.

Dams are often built across rivers to hold back large amounts of water. When the water is released, it flows through pipes inside the dam and turns turbines (which are like big wheels). These turbines are connected to a generator, which makes electricity that can power homes, schools, and cities!

Activity

Make a Water Wheel:

Materials: A plastic cup, cardboard, a wooden skewer, tape, and scissors.

Instructions:

1. Cut the cardboard into strips and bend them to make small paddles.
2. Attach the paddles to the outside of the plastic cup to create a water wheel.
3. Push the skewer through the middle of the cup, allowing the wheel to spin.
4. Hold the skewer over a bowl and pour water over the paddles to see the wheel turn!

Learning Moment: This water wheel is similar to how a dam's turbines spin when water flows over them, creating energy.

Biomass: Energy from Nature's Leftovers

Biomass is a type of energy that comes from natural materials like plants, trees, and even food waste! Instead of letting things like wood scraps, fallen leaves, or leftover food go to waste, we can use them to make fuel. This fuel can be burned to produce heat and electricity, or even turned into gas and liquid fuels to power cars and other machines.

How it Works

When organic materials are burned, they release energy in the form of heat. This heat can be used directly, or it can be converted into electricity. In some cases, the organic material is broken down in special machines to create biogas, which can be used to cook or heat homes.

Activity
Make a Mini Compost Jar

Materials: A clear jar, soil, leaves, vegetable scraps, and a lid with small holes.

Instructions:

1. Place a layer of soil at the bottom of the jar.
2. Add a small layer of leaves, followed by some vegetable scraps.
3. Repeat these layers until the jar is full, ending with a layer of soil on top.
4. Put the lid on and let it sit in a warm place. Over time, the materials will break down and turn into compost!

Learning Moment: The materials break down, they release energy in the form of heat, similar to how biomass creates energy. Plus, kids can see the natural process of decomposition!

Geothermal Energy: Heat from the Earth

Deep beneath the Earth's surface, it's really hot! This heat is called geothermal energy. We can use this natural heat to warm up our homes, make hot water, and even generate electricity.

Geothermal energy comes from hot rocks and underground reservoirs of steam or hot water. To harness this energy, we drill wells deep into the Earth to reach the hot water. The hot water or steam is then pumped up to the surface, where it can turn a turbine in a power plant to create electricity. After it's used, the water goes back underground, where it gets heated again by the Earth.

Activity

Hot Rocks Experiment

Materials: A clear glass bowl, hot water, small stones, and a thermometer.

Instructions

1. Place the stones in the glass bowl.
2. Pour hot water over the stones, then measure the temperature of the water.
3. Watch as the stones heat up and help keep the water warm.

Learning Moment: The hot rocks under the Earth, these stones help hold the heat, which is similar to how geothermal energy works.

Tidal Energy: Power from the Ocean's Waves

The ocean's tides move in and out every day, thanks to the gravitational pull of the moon and the sun. Tidal energy captures this powerful movement of water to make electricity. Tides are very predictable, which makes tidal energy a reliable source of power!

How it Works

Special machines called tidal turbines or tidal barrages are placed in the ocean. As the tide comes in and goes out, it pushes water through these machines. The moving water spins the turbines, which are connected to a generator that creates electricity. It's similar to how wind turbines work, but underwater!

Activity

Tide Observation

Materials: A notebook, pen, and a visit to a beach or online tide chart.

Instructions:

- Observe or look up the tide schedule for a local beach.

- Record the high and low tides and note how the water level changes throughout the day.

Generate power, similar to how they see waves moving in and out.

Wave Energy: Power from Ocean Waves

The ocean's waves are always in motion, carrying a lot of energy. Wave energy captures this movement and turns it into electricity. It's similar to tidal energy, but instead of using the tide moving in and out, it uses the up-and-down motion of the waves.

Special machines called wave energy converters are placed on the surface of the ocean or just below it. As the waves pass, they push these machines up and down, which moves parts inside the machines and creates electricity. Some wave converters look like floating buoys, while others are anchored to the ocean floor

Activity

Observe Waves at the Beach:

Materials: A small toy boat or a floating object, a bowl of water, and your hand.

Instructions

Fill the bowl with water and place the toy boat on top.

Use your hand to create small waves by gently pushing the water.

Watch as the toy boat moves up and down with the waves.

Learning Moment: The boat's movement is similar to how wave converters work. As the waves move, the converters bob up and down, creating energy.

Non-Renewable Energy

Non-Renewable Energy comes from sources that can run out because they take millions of years to form. Once we use them up, they're gone for good!

Non-renewable energy sources include:

Fossil Fuels

Coal: Formed from ancient plants buried under layers of earth and water millions of years ago. When we burn coal for energy, it releases smoke and pollutants into the air.

Oil: Created from ancient sea creatures that decomposed over millions of years under the earth. Oil is refined to make gasoline, diesel, and other fuels.

Natural Gas: Also formed from decomposed plants and animals. It's often used for heating homes, cooking, and generating electricity.

2. Nuclear Energy:

Nuclear energy uses uranium, a mineral found in the earth, to create electricity. The process involves splitting uranium atoms to release large amounts of energy. Although it doesn't produce air pollution, it creates radioactive waste, which is challenging to safely store.

Why is Non-Renewable Energy a Problem?

Using non-renewable energy sources has some downsides:

Pollution: Burning fossil fuels releases harmful gases that contribute to air pollution and climate change.

Finite Resources: Once these resources are used up, they're gone forever. We can't make more coal, oil, or gas within a human lifetime.

Environmental Impact: Extracting these resources can damage ecosystems, such as drilling for oil or mining for coal, which affects animals and plants.

Fun Fact:

Did you know that over 80% of the world's energy still comes from non-renewable sources? That's why many countries are working hard to switch to renewable energy!

Questions

1. What is renewable energy, and why is it important?

2. Can you name three types of renewable energy?

3. How does solar energy work, and what are solar panels used for?

4. Why do you think wind turbines are usually placed in open fields or on hills?

5. What materials are used to create biomass energy? Can you think of any examples?

6. Where does geothermal energy come from, and how do we use it?

7. How do the tides help us create electricity?

8. What is the difference between tidal energy and wave energy?

9. Can you list some of the benefits of using renewable energy instead of non-renewable energy?

10. Which type of renewable energy do you find the most interesting, and why?

11. If you could use one type of renewable energy in your home, which one would you choose?

12. How do you think renewable energy can help protect animals and the environment?

13. Imagine you're an engineer. What kind of renewable energy invention would you like to create?

14. How can you help your family or school use more renewable energy in everyday life?

Non-Renewable Energy Questions

1. What are non-renewable energy sources? Can you name two examples?

2. Why are coal, oil, and natural gas called fossil fuels?

3. How do fossil fuels form, and why does it take millions of years?

4. What happens when we burn fossil fuels to make electricity?

5. What are some problems with using non-renewable energy sources?

6. How is nuclear energy different from fossil fuels?

7. Why can't we use non-renewable energy sources forever?

8. What is one way that using fossil fuels can harm the environment?

9. How is pollution from non-renewable energy different from renewable energy?

10. What are some things we can do to use less non-renewable energy at home or school?

11. Can you think of any ways that fossil fuels impact animals and nature?

12. Why do you think it's important to start using more renewable energy instead of fossil fuels?

13. What would happen if we ran out of non-renewable energy sources?

14. What kinds of activities rely on non-renewable energy sources?

15. If you were in charge of energy for your city, would you choose non-renewable or renewable sources, and why?

The Sun Powers More than You Think! Every hour, the sun sends more energy to Earth than the entire world uses in a year!

Wind Turbines are Huge! The blades of a typical wind turbine are as long as a Boeing 747's wingspan. The taller the turbine, the more wind it catches.

Geothermal Energy is Really Old. People have used geothermal energy for thousands of years; ancient Romans used hot springs to heat their homes.

Biomass Includes Leftovers. Biomass energy can come from food scraps, fallen leaves, and sawdust, turning waste into energy and reducing garbage.

Ocean Waves are Powerful. Wave energy could potentially power millions of homes, even in rough seas, because of the ocean's strong motion.